OFFICIAL DISCARD
LaGrange County Public Library

MONSTER MACHINES

Paul Nash

Some of the biggest machines on earth for lifting, cutting, drilling, sailing, unloading, transporting, squirting, generating, stacking, hovering, dredging, harvesting, flying — and talking!

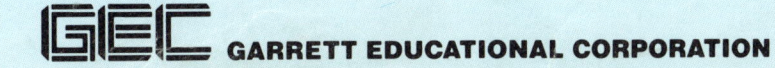

GEC GARRETT EDUCATIONAL CORPORATION

CONTENTS and Picture Acknowledgments

Mining dump truck	3
Caterpillar Tractor Company, Peoria, Illinois	
Truck-mounted crane	4
G.W. Sparrow & Sons, Bath, England	
Offshore production platform	6
The British Petroleum Company Ltd., London, England	
Combine harvester	7
Massey-Ferguson (United Kingdom) Ltd., Warwickshire, England	
Walking dragline	8
Bucyrus-Erie, South Milwaukee, Wisconsin	
Articulated dump truck	10
DJB Engineering Ltd., County Durham, England	
Ship unloaders	11
Simon Engineering Ltd., Cheshire, England	
Passenger hovercraft	12
British Hovercraft Corporation, Isle of Wight, England	
Mining tractor	13
Caterpillar Tractor Company, Peoria, Illinois	
Coal dredger	14
State Electricity Commission of Victoria, Australia	
Semi-submersible crane vessel	16
Heerema Engineering Service B.V., Leiden, The Netherlands	
Electricity turbogenerator	19
NEI Parsons Ltd., Newcastle upon Tyne, England; and Central Electricity Generating Board, London, England	
Boom-stacker for coal	20
Central Electricity Generating Board, London, England	
Longest bus in world	22
Wayne Corporation, Richmond, Indiana	
Railway track foundation laying machine	23
British Railways Board, London, England	
Rocket transporter	24
N.A.S.A., Cape Canaveral, Florida	
Aircraft carrier	25
Department of the Navy, Washington, D.C.	
Offshore fire-fighting vessel	26
The British Petroleum Company Ltd., London, England	
Big Shovel	29
Peabody Holding Company Inc., St. Louis, Missouri	
Passenger aircraft	30
Boeing Commercial Airplane Company, Seattle, Washington	
Telephone	32
British Telecom, London, England	

Text © copyright 1989 by Garrett Educational Corporation
First published in the United States in 1989 by Garrett Educational Corporation, 130 East 13th Street, Ada, OK 74820.
First Published by Young Library Ltd., Brighton, England
© Copyright 1983 Young Library Ltd.

All rights reserved including the right of reproduction in whole or in part in any form without the prior written permission of the publisher. Published by Garrett Educational Corporation, 130 East 13th Street, P.O. Box 1588, Ada, Oklahoma 74820.

Manufactured in the United States of America

Library of Congress Cataloging in Publication Data

Nash, Paul, 1943-
  Monster machines / Paul Nash.
    p. cm.
  Summary: An introduction to some of the biggest machines on earth.
  1. Machinery—Juvenile literature. [1. Machinery.]
I. Title.
TJ147.N37   1989
621.8—dc20   89-12010
  ISBN 0-944483-36-4   CIP
                                              AC

MINING DUMP TRUCK

This Caterpillar dump truck can move 76 tons (77,000 kgs) of rocks and earth at 37 mph (60 km/h) over fairly bumpy ground. The V-type 12-cylinder diesel engine has power equal to 11 cars, with seven forward gears and one reverse gear. The fuel tank holds enough to fill the tanks of 21 family cars.

kgs: kilograms mass or weight (10 kgs = 22 pounds).
km/h: speed in kilometers an hour.
mph: speed in miles per hour (10 mph = 16 km/h).
gears: a mechanism for changing the speed of an engine.

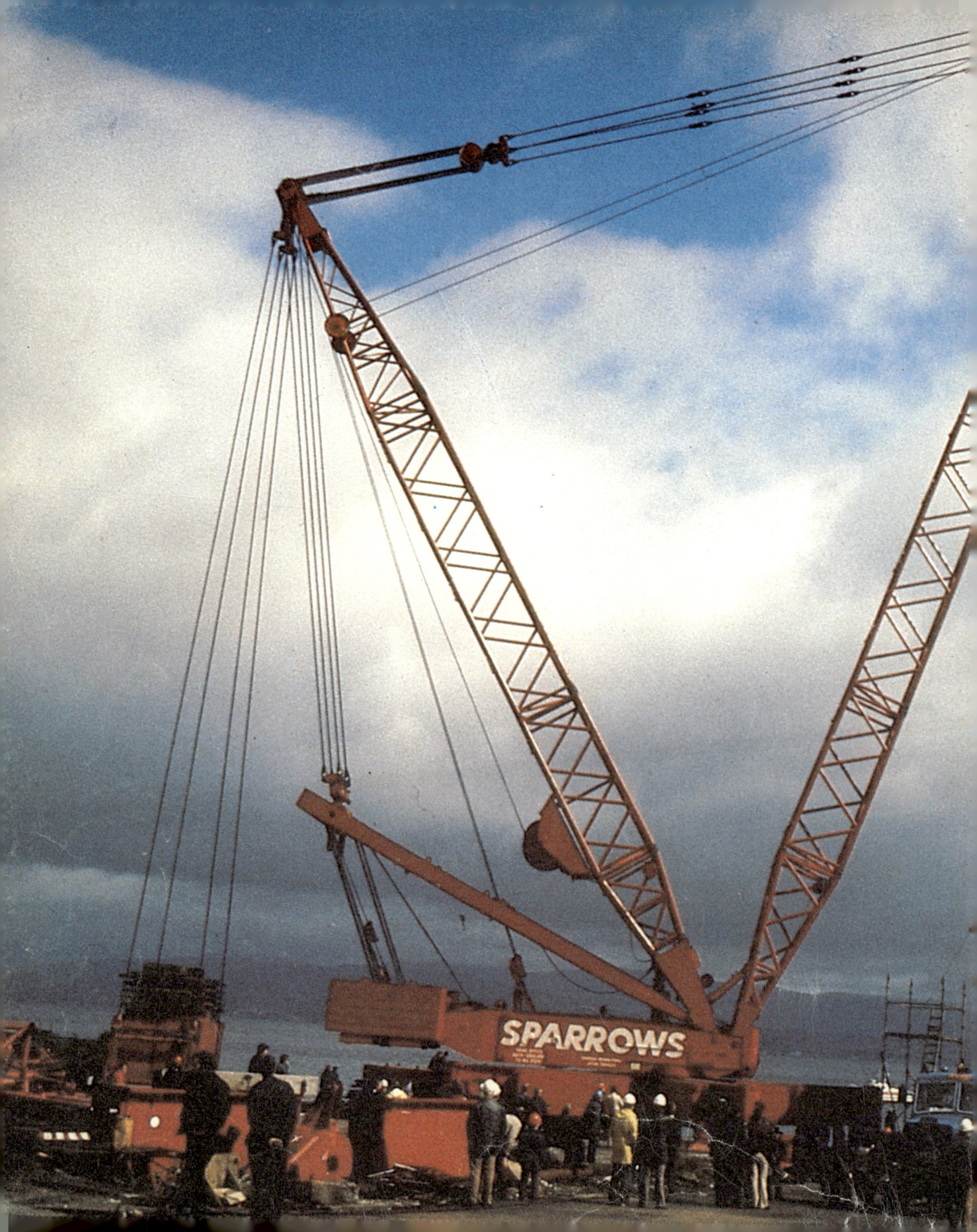
SPARROWS

TRUCK-MOUNTED CRANE

Here is the world's largest truck-mounted crane launching a new 98-foot (30-meter) ship. The MK1000 Gottwald crane can lift up to 1,020 tons (1,000,000 kgs) — that's the same weight as 146 elephants. It has a 167-foot (50-meter) main boom and can reach out to 66 feet (20 meters) when moving heavy loads from one place to another.

boom: a rigid steel structure like an arm which allows a machine to reach out to do its work.

OFFSHORE PRODUCTION PLATFORM

This is a bird's-eye view of a massive oil production platform in the North Sea. It stands 1,020 feet (312 meters) high, twice as tall as the Post Office Tower in London. You can see several cranes, and at the top of the picture is a helicopter pad. Each of the legs of this production platform is bigger than any other steel structure in the world. Oil from the well is pumped at the rate of 50 gallons (220 liters) a second to the mainland 125 miles (200 kilometers) away.

COMBINE HARVESTER

This powerful grain harvester cuts down wheat stalks and separates the grain which will be made into bread flour. It produces grain sufficient for six large loaves of bread a second — 10.2 tons (10,000 kgs) an hour. While traveling across the field at a fast walking pace with its 20-foot (6-meter) cutter — as wide as a house — it harvests an area the size of a football field in just over half an hour.

WALKING DRAGLINE

This is believed to be the largest mobile land machine in the world. The "Big Muskie" Bucyrus-Erie 4250-W walking dragline excavator has a total weight of 12,000 tons (12,200,000 kgs). It takes a massive 197-ton (200,000-kg) bite as its excavator bucket is dragged over the surface of the open-pit coal mine where it is used. A small cottage could fit comfortably inside the 220-cubic-yard) 170-cubic-meter bucket. To help it reach out, the dragline has a boom nearly as long as a football field, along which the bucket is pulled. Caterpillar tracks enable the machine to move about.

caterpillar tracks: an endless metal chain fastened over the wheels of a vehicle to enable it to move over soft or uneven ground; as the wheels turn around, the track-slabs revolve around them so that the wheels do not touch the ground.

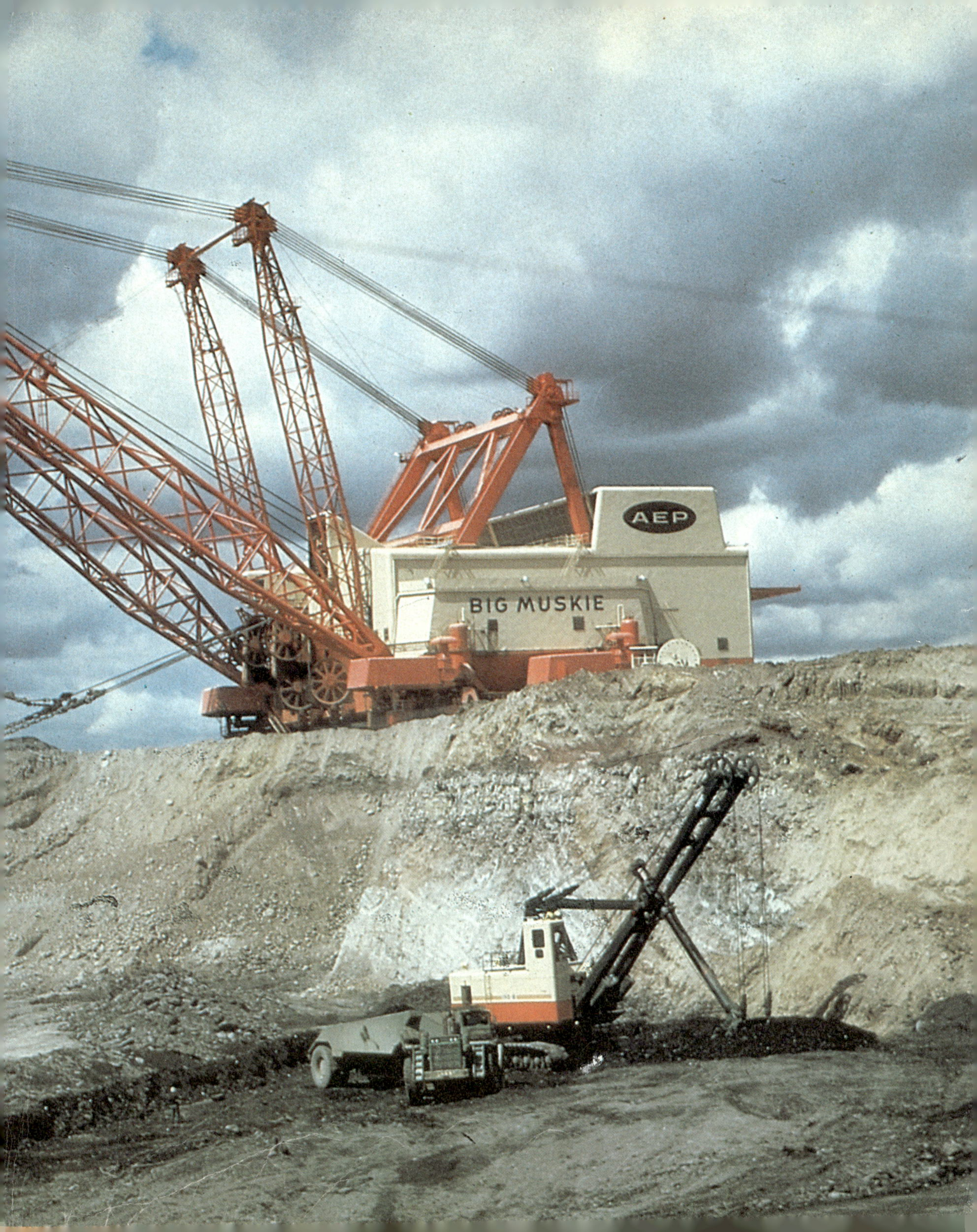
AEP
BIG MUSKIE

ARTICULATED DUMP TRUCK

This D550 articulated truck is specially designed to turn around tight corners in crowded mines. It is ruggedly built for rough work, and is the largest in the world. It has a pivot to allow the front cab with its 450-horsepower engine to move at an angle (or "articulate") in relation to the rear load-carrying body, which weighs 51 tons (50,000 kgs). The turning circle for this 37-foot- (11-meter-) long truck is a mere 63 feet (19 meters), not much when you think a passenger car needs 33 feet (10 meters) in which to turn around.

articulated: front part (cab and engine) of a truck being able to turn at an angle to the rear load, instead of being rigidly attached in line.
turning circle: minimum space needed to drive a vehicle completely around in a circle.

SHIP UNLOADERS

These two machines stand 50 feet (15 meters) high and unload grain from the cargo-holds of ships at the rate of 740 tons (750,000 kgs) an hour each. Together they move 920 pounds (420 kgs) every second they are working. To do the same job at that speed, you would need 600 men rushing around working as fast as possible. As well as doing the unloading quickly, the unloaders are gentle — they carry the food grains between two moving belts which are held together by air pressure, so that the grains are not damaged.

PASSENGER HOVERCRAFT

This B.H.C. Super-4 Hovercraft, invented and built in England, is the world's largest vessel of its type. It can carry 416 passengers and 55 cars at 75 mph (120 km/h), which is much faster than any ship. Gliding over the waves on a cushion of air, it's more comfortable, too! Each of the four propellers, as high as a house, is driven by a Rolls-Royce 3,800-horsepower gas-turbine engine. Five Super-4 hovercraft, each 185 feet (56 meters) long and 92 feet (28 meters) wide, would fill the average-size football field.

turbine: an engine driven by air, steam, hot gases, or water pushing fan-blades around.

MINING TRACTOR

This D10 track-type tractor was designed by Caterpillar Tractor Co. to work in open mines. With its 20-foot (6-meter) bulldozer blade it can push 34 cubic yards (25 cubic meters) — that's two normal truckfuls — at a time. The driver sits 12 feet (3.5 meters) above the ground, which is like sitting in an upstairs bedroom. The 700-horsepower, 29-liter, 12-cylinder diesel engine is nine times more powerful than an average car engine. It is not very fast, however. Top speed is in reverse third gear, 9 mph (14 km/h) — jogging speed. From a standing start, the tractor can pull 123 tons (125,000 kgs); that's more than its own weight of 93 tons (94,700 kgs).

COAL DREDGER

This is one of several giant coal dredgers that cut soft brown coal in an open-pit mine in Australia. The biggest machine like this digs out 3,760 tons (3,700,000 kgs) an hour; that's 21 sacks (each 112 pounds, 50 kgs) every second. The ten buckets at the front of the dredger are each the size of a small delivery van (450 gallons, 2 cubic meters), and scoop out coal at the rate of 80 giant bucketfuls a minute. The dredger stands as high as the seventeenth floor of an apartment building, and can move to its next workplace at 26 feet (8 meters) a minute — just twice as fast as a turtle!

open-pit mine: a mine that is made by digging at the surface rather than tunneling below it.

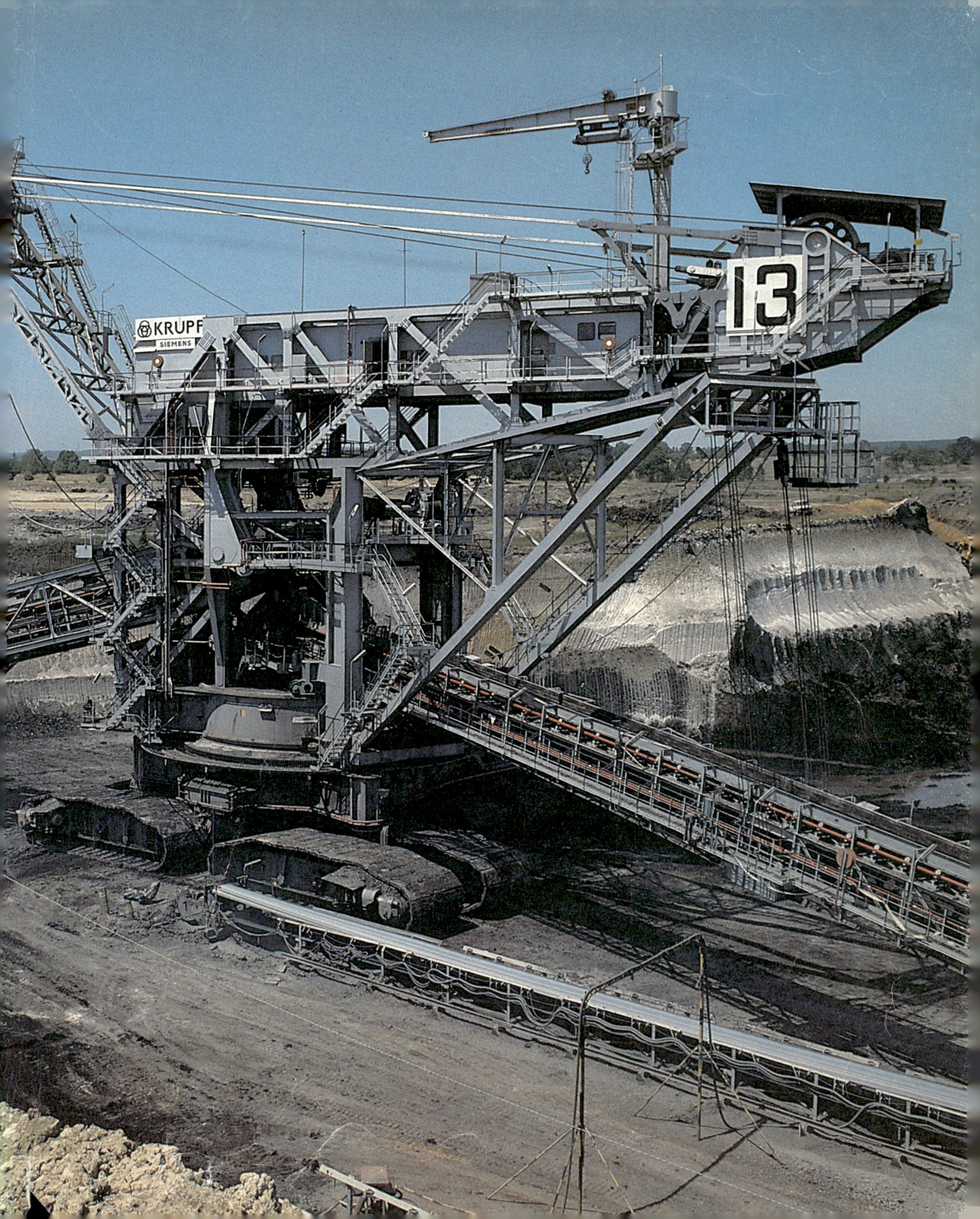
KRUPP
SIEMENS
13

SEMI-SUBMERSIBLE CRANE VESSEL

This is the largest semi-submersible crane vessel in the world. One of its giant cranes can lift 1,970 tons (2,000,000 kgs) on to an oil rig, the other can lift 2,950 tons (3,000,000 kgs). It has living quarters for 550 men, and is propelled to its work site at sea by its own 16,000-horsepower engines, equivalent to 200 cars. The vessel can adjust its deck height to make itself more stable in rough weather. This up-and-down adjustment from 30-90 feet (9 to 27.5 meters) above sea level is done by varying its buoyancy. Once correctly positioned, the floating vessel is fixed with 16 anchors.

buoyancy: the power to float or rise in water.
semi-submersible: a vessel that can change the proportion of its structure that is under water.

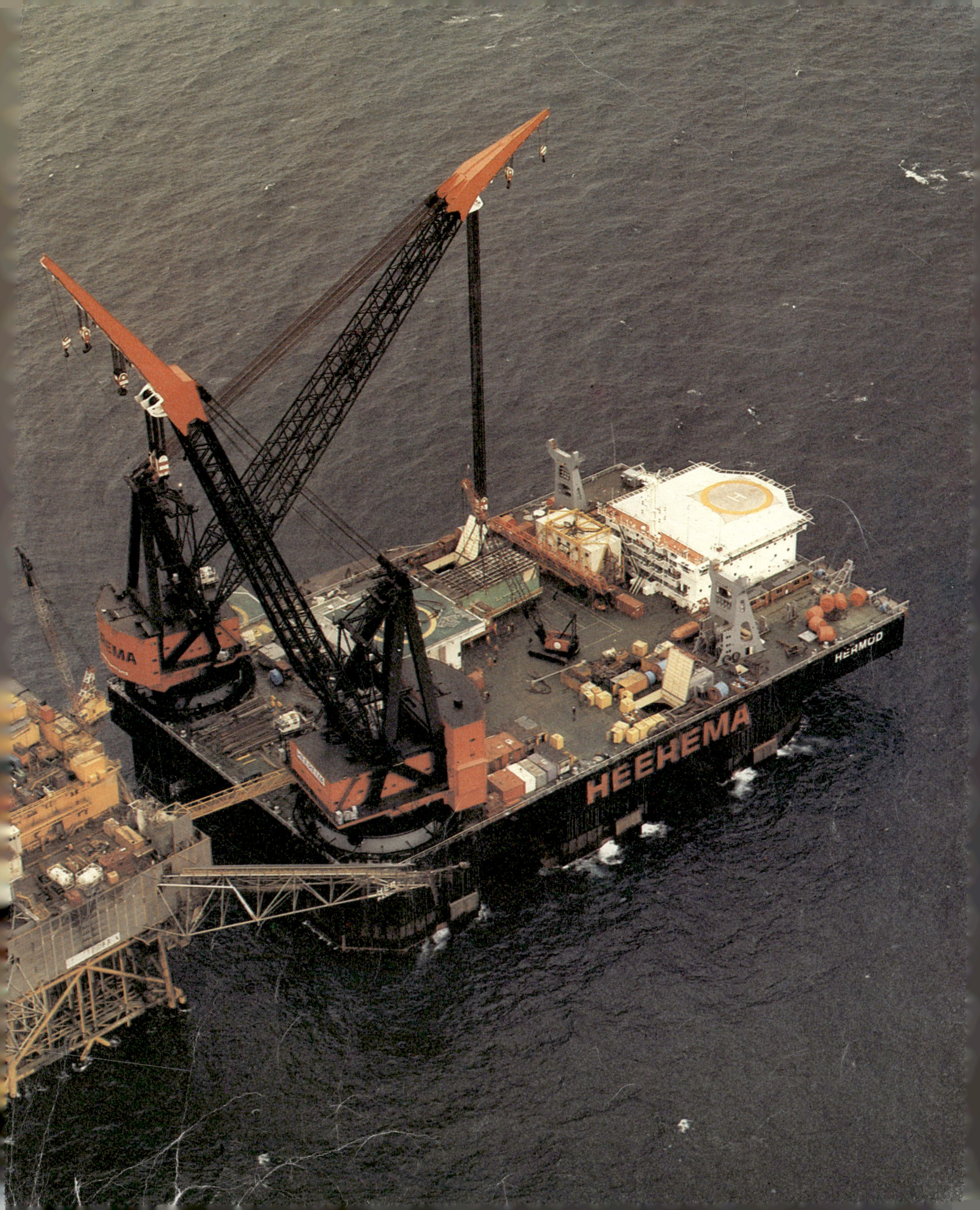
HEEREMA
HERMOD

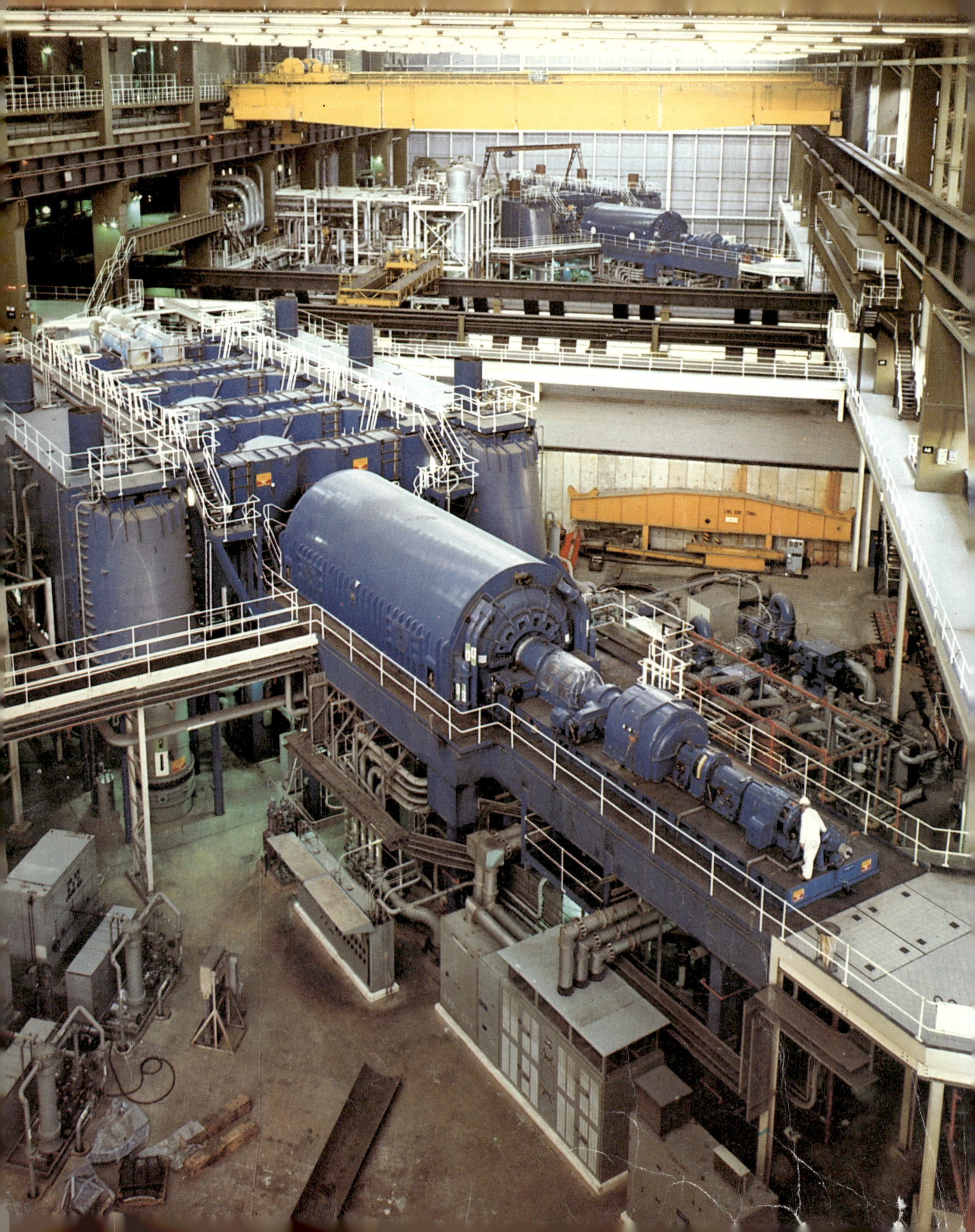

ELECTRICITY TURBOGENERATOR

Britain's largest power station at Drax in North Yorkshire has three giant turbogenerators. Each generates enough electricity (660 MW) to run 330,000 two-burner electric stoves. The rotor of the turbine, which you can see in the smaller picture, has a total of 14,400 blades that are pushed round by high-pressure steam from the coal-fired boiler. Heat output from that boiler is equivalent to boiling ten bathfuls (each 45 gallons, 200 liters) every second.

MW: megawatts; 1 MW = 1,000,000 watts, equivalent to 1,000 one-burner electric stoves.
turbogenerator: a machine for producing electricity driven by a steam turbine.

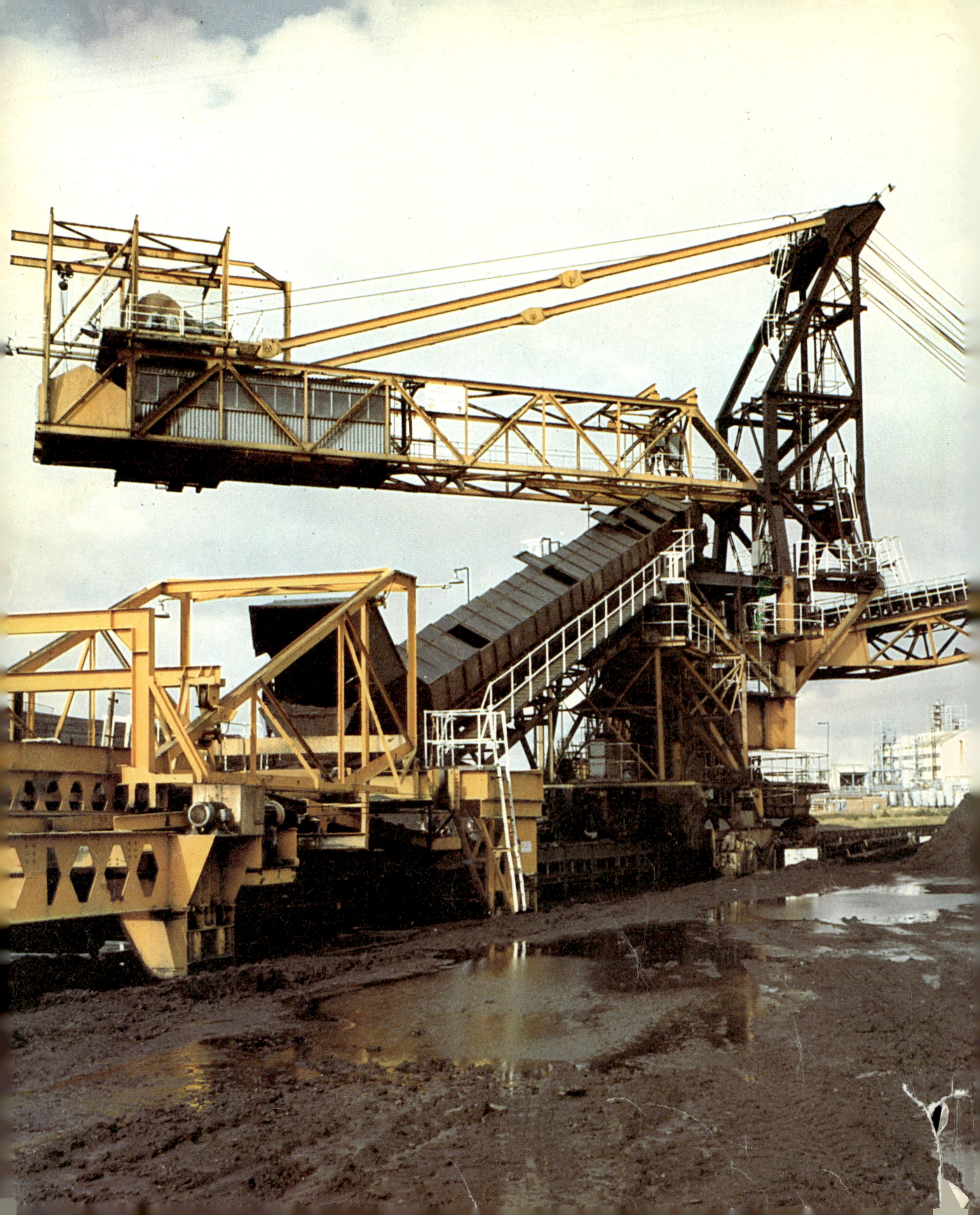

BOOM-STACKER FOR COAL

This giant stacker works at a power station where the coal is delivered by 20 railway trains a day, each carrying 1,000 tons (1,020,000 kgs) of coal. As the train moves slowly forward at 0.5 mph (0.8 km/h), the coal is automatically discharged onto high-speed conveyors. The boom-stacker shown here can pile up the coal from these conveyors into big storage stacks at the rate of 3,000 tons (3,050,000 kgs) an hour — the same as 17 bags (each 112 pounds, 50 kgs) every second. Digging out this coal, the bucket-wheel at the front scoops it up at the rate of 1,500 tons (1,520,000 kgs) an hour — which is the same as 140 men lifting one shovelful every second!

conveyor: an endless belt to take materials from one place to another, the empty belt returning in the opposite direction underneath the full belt.

LONGEST BUS IN WORLD

To take people to and from their workplace in the Middle East oilfields, this giant bus can move up to 187 passengers at once — 121 sitting and 66 standing. At 76 feet (23 meters), it is as long as a tennis court. A separate power unit tows the body, which weighs 10.8 tons (11,000 kilograms), and can carry a human cargo weighing 14.8 tons (15,000 kgs).

RAILWAY TRACK FOUNDATION LAYING MACHINE

This machine continually casts a reinforced concrete foundation 1 foot (30 centimeters) thick for railway tracks. It is made up of five parts, and is 200 feet (60 meters) in length. The paving machine moves along at a snail's pace, 130 feet (40 meters) an hour. Every ten minutes a truckload of concrete is emptied into the front elevator. A conveyor takes the concrete to the rear of the machine. The next two gantries lay down the reinforcing rods, which are then covered with concrete by the fourth machine. Lastly, holes are blown in the wet concrete by air blasts to allow anchoring bolts to be set in.

foundation: the base on which a building, machine, or railway track is supported.
gantries: mobile overhead frameworks.

ROCKET TRANSPORTER

The most massive transporters in the world are used at the John F. Kennedy Space Center in Florida to move the huge Saturn-V space rockets from their assembly building to the launch pad. The rocket stands 364 feet (110 meters) high, and sits on the transporter that is big enough (131 × 114 feet, 40 × 35 meters) to cover five tennis courts. Fully loaded, the eight-track caterpillar crawler weighs 8,040 tons (8,170,000 kgs) and needs 42 inch (110 centimeter) wiper blades for its windshields.

AIRCRAFT CARRIER

The nuclear-powered U.S.S. *Dwight D. Eisenhower* is the largest aircraft carrier in the world. It weighs 91,400 tons (92,900,000 kgs) when fully loaded with its crew of 6,100. The 90 jet planes on board have 1,090 feet (332 meters) in which to take off — a tiny bit shorter than the height of the Empire State Building. The nuclear reactors which power the 280,000-horsepower turboengines can go 900,000 miles (1,450,000 kilometers) before needing to refuel.

nuclear (power, or reactor): a way of generating power by splitting atoms.
U.S.S.: United States Ship.

OFFSHORE FIRE-FIGHTING VESSEL

The Emergency Support Vessel *Iolair* is designed to fight fires if they break out on the offshore oil rigs. It is used by British Petroleum in the North Sea Forties oilfield. Here is *Iolair* showing off its four big water jets, which together can throw seawater over 590 feet (180 meters) at a rate of 3,200,000 gallons (14,400,000 liters) an hour — that's 20 bathfuls of water every second. Seawater is sucked up and squirted by very powerful pumps whose combined power (27,000-horsepower) is that of 330 car engines. Seawater is also sprayed down *Iolair's* sides to keep the vessel cool and allow it to fight fire right alongside a burning oil rig.

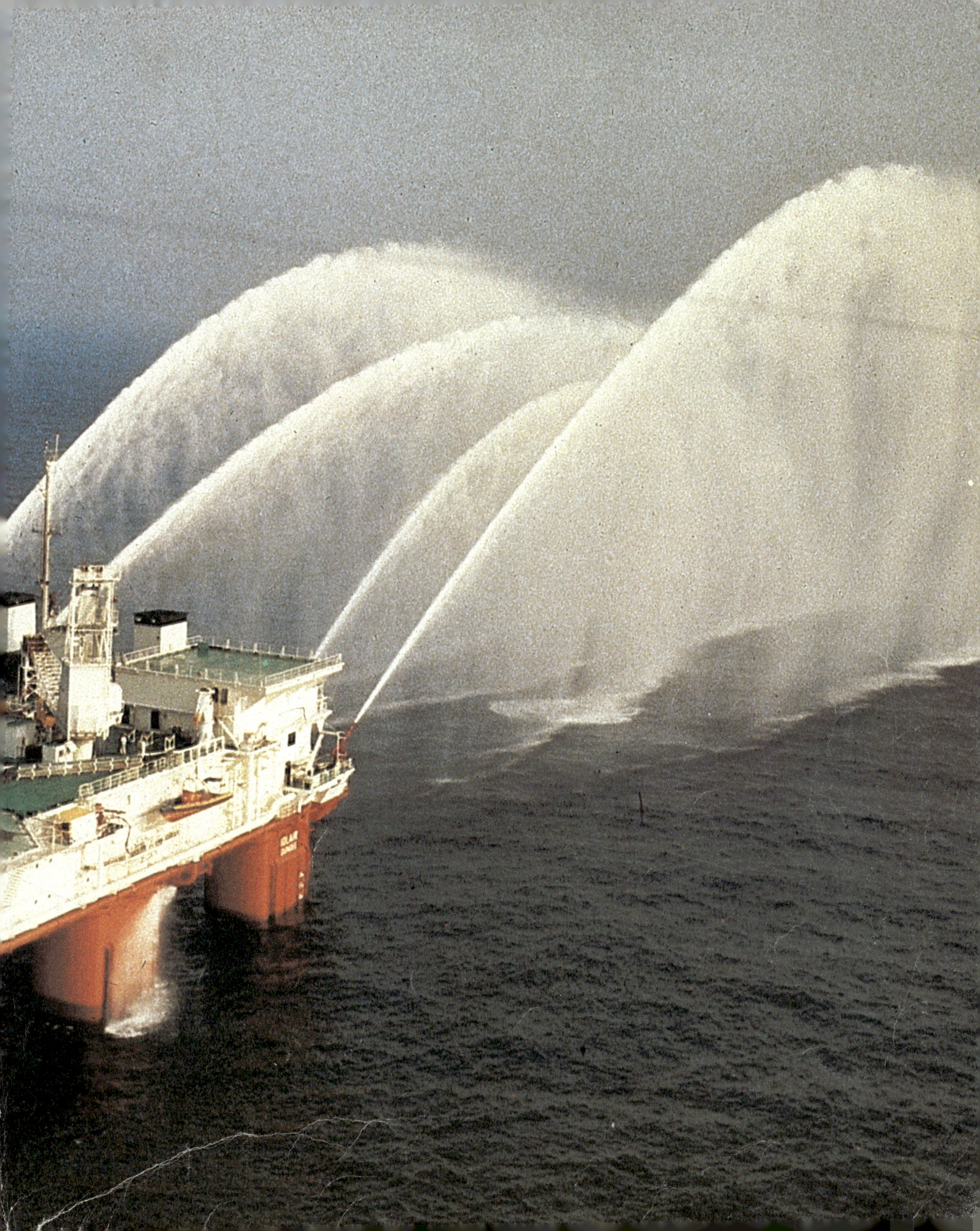

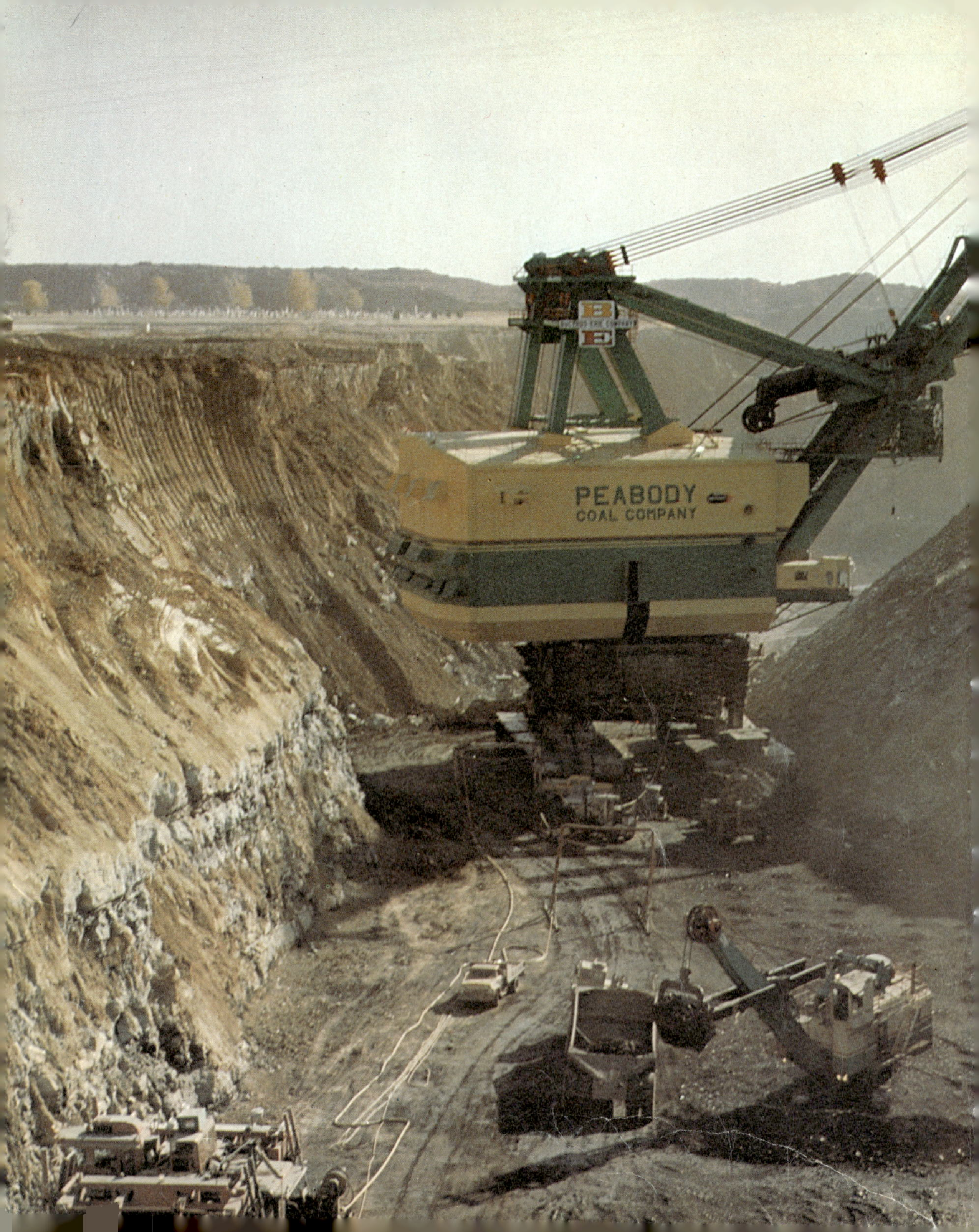
PEABODY
COAL COMPANY

BIG SHOVEL

This monster shovel works at an open-pit coal mine in the United States, shoveling up coal at 13,000 tons (14,000,000 kgs) a day — that's 8 sackfuls (each 112 pounds, 50 kgs) every second. As high as a 20-story apartment house (250 feet, 76 meters), and as wide as an eight-lane highway, it is powered by electricity from the 7,200-volt cable that is trailed behind. The shovel removes enough earth and coal each month (3,330,000 cubic yards, 2,580,000 cubic meters) to make one of Egypt's ancient pyramids every month.

PASSENGER AIRCRAFT

Over 450 passengers can fly up to 5,500 miles (8,800 kilometers) in comfort in the Boeing 747, the largest passenger plane in the world. The fuselage is 232 feet (71 meters) long, and the wing tips nearly 196 feet (60 meters) apart, two-thirds the length of a football field.

Each of the four jets generates 54,000 pounds (24,500 kgs) of thrust, which is like the weight of eight elephants pushing hard. Enough fuel is used on the journey between Los Angeles and London to fill the gasoline tank of a car once a week for 70 years.

fuselage: the main body of a plane where passengers and cargo are carried.
thrust: the force with which a jet engine pushes the air behind it.

And on the next page
is a skilled operator
at one of the terminals
of by far the largest machine
in the entire world . . .

31

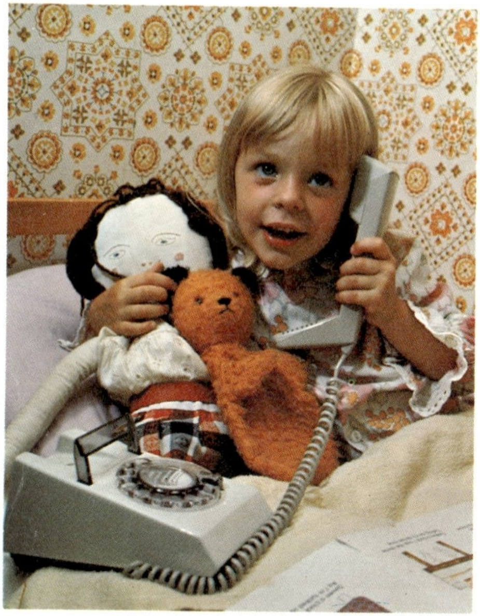